# ENERGY SECTOR STANDARD OF THE PEOPLE'S REPUBLIC OF CHINA

中华人民共和国能源行业标准

## Specification for Preparation of Flood Control Report During Construction Period of Hydropower Projects

水电工程施工期防洪度汛报告编制规程

NB/T 10492-2021

Chief Development Department: China Renewable Energy Engineering Institute
Approval Department: National Energy Administration of the People's Republic of China
Implementation Date: July 1, 2021

China Water & Power Press

Beijing 2024

All rights reserved. No part of this publication may be reproduced, stored in a retrieval system, or transmitted in any form or by any means—electronic, mechanical, photocopying, recording or otherwise, without prior written permission of the publisher.

图书在版编目（CIP）数据

水电工程施工期防洪度汛报告编制规程：NB/T 10492-2021 = Specification for Preparation of Flood Control Report During Construction Period of Hydropower Projects (NB/T 10492-2021)：英文 / 国家能源局发布. -- 北京：中国水利水电出版社，2024. 8. -- ISBN 978-7-5226-2665-9

Ⅰ. TV87

中国国家版本馆CIP数据核字第2024KJ9482号

ENERGY SECTOR STANDARD
OF THE PEOPLE'S REPUBLIC OF CHINA
中华人民共和国能源行业标准

Specification for Preparation of Flood Control Report
During Construction Period of Hydropower Projects
水电工程施工期防洪度汛报告编制规程
NB/T 10492-2021
（英文版）

Issued by National Energy Administration of the People's Republic of China
国家能源局　发布
Translation organized by China Renewable Energy Engineering Institute
水电水利规划设计总院　组织翻译
Published by China Water & Power Press
中国水利水电出版社　出版发行
　　Tel:（+ 86 10）68545888　68545874
　　sales@mwr.gov.cn
　　Account name: China Water & Power Press
　　Address: No.1, Yuyuantan Nanlu, Haidian District, Beijing 100038, China
　　http://www.waterpub.com.cn
中国水利水电出版社微机排版中心　排版
北京中献拓方科技发展有限公司　印刷
184mm×260mm　16开本　2.25印张　71千字
2024年8月第1版　2024年8月第1次印刷
**Price**（定价）：￥360.00

# Introduction

This English version is one of China's energy sector standard series in English. Its translation was organized by China Renewable Energy Engineering Institute authorized by National Energy Administration of the People's Republic of China in compliance with relevant procedures and stipulations. This English version was issued by National Energy Administration of the People's Republic of China in Announcement [2023] No. 8 dated December 28, 2023.

This version was translated from the Chinese Standard NB/T 10492-2021, *Specification for Preparation of Flood Control Report During Construction Period of Hydropower Projects*, published by China Water & Power Press. The copyright is reserved by National Energy Administration of the People's Republic of China. In the event of any discrepancy in the implementation, the Chinese version shall prevail.

Many thanks go to the staff from the relevant standard development organizations and those who have provided generous assistance in the translation and review process.

For further improvement of the English version, any comments and suggestions are welcome and should be addressed to:

China Renewable Energy Engineering Institute
No. 2 Beixiaojie, Liupukang, Xicheng District, Beijing 100120, China
Website: www.creei.cn

Translating organizations:

POWERCHINA Guiyang Engineering Corporation Limited

China Renewable Energy Engineering Institute

Translating staff:

| | | | |
|---|---|---|---|
| HAO Peng | SONG Tao | WANG Xiangyu | LIU Ming |
| ZHENG Qing'an | WEI Fang | GUO Yong | GE Xiaobo |

Review panel members:

| | |
|---|---|
| LIU Xiaofen | POWERCHINA Zhongnan Engineering Corporation Limited |
| YAN Wenjun | Army Academy of Armored Forces, PLA |
| LI Zhongjie | POWERCHINA Northwest Engineering Corporation Limited |

| | |
|---|---|
| CHEN Lei | POWERCHINA Zhongnan Engineering Corporation Limited |
| QIAO Peng | POWERCHINA Northwest Engineering Corporation Limited |
| YE Bin | POWERCHINA Huadong Engineering Corporation Limited |
| XU Zeping | China Institute of Water Resources and Hydropower Research |
| LI Shisheng | China Renewable Energy Engineering Institute |

National Energy Administration of the People's Republic of China

# 翻译出版说明

本译本为国家能源局委托水电水利规划设计总院按照有关程序和规定，统一组织翻译的能源行业标准英文版系列译本之一。2023年12月28日，国家能源局以2023年第8号公告予以公布。

本译本是根据中国水利水电出版社出版的《水电工程施工期防洪度汛报告编制规程》NB/T 10492—2021翻译的，著作权归国家能源局所有。在使用过程中，如出现异议，以中文版为准。

本译本在翻译和审核过程中，本标准编制单位及编制组有关成员给予了积极协助。

为不断提高本译本的质量，欢迎使用者提出意见和建议，并反馈给水电水利规划设计总院。

  地址：北京市西城区六铺炕北小街2号
  邮编：100120
  网址：www.creei.cn

本译本翻译单位：中国电建集团贵阳勘测设计研究院有限公司
       水电水利规划设计总院
本译本翻译人员：郝 鹏 宋 涛 王向予 刘 明
       郑庆安 魏 芳 郭 勇 葛小博
本译本审核人员：
  刘小芬 中国电建集团中南勘测设计研究院有限公司
  闫文军 中国人民解放军陆军装甲兵学院
  李仲杰 中国电建集团西北勘测设计研究院有限公司
  陈 蕾 中国电建集团中南勘测设计研究院有限公司
  乔 鹏 中国电建集团西北勘测设计研究院有限公司
  叶 彬 中国电建集团华东勘测设计研究院有限公司
  徐泽平 中国水利水电科学研究院
  李仕胜 水电水利规划设计总院

国家能源局

# Announcement of National Energy Administration of the People's Republic of China [2021] No. 1

National Energy Administration of the People's Republic of China has approved and issued 320 energy sector standards including *Code for Integrated Resettlement Design of Hydropower Projects* (Attachment 1), the foreign language versions of 113 energy sector standards including *Carbon Steel and Low Alloy Steel for Pressurized Water Reactor Nuclear Power Plants—Part 7: Class 1, 2, 3 Plates* (Attachment 2), and the amendment notification for 5 energy sector standards including *Technical Code for Investigation and Assessment of Aquatic Ecosystem for Hydropower Projects* (Attachment 3).

Attachments: 1. Directory of Sector Standards

2. Directory of Foreign Language Versions of Sector Standards

3. Amendment Notification of Sector Standards

National Energy Administration for the People's Republic of China

January 7, 2021

Attachment 1:

**Directory of Sector Standards**

| Serial number | Standard No. | Title | Replaced standard No. | Adopted international standard No. | Approval date | Implementation date |
|---|---|---|---|---|---|---|
| ... | | | | | | |
| 9 | NB/T 10492-2021 | Specification for Preparation of Flood Control Report During Construction Period of Hydropower Projects | | | 2021-01-07 | 2021-07-01 |
| ... | | | | | | |

# Foreword

According to the requirements of Document GNKJ [2016] No. 238 issued by National Energy Administration of the People's Republic of China, "Notice on Releasing the Development and Revision Plan of Energy Sector Standards in 2016", and after extensive investigation and research, summarization of practical experience, and wide solicitation of opinions, the drafting group has prepared this specification.

The main technical contents of this specification include: general provisions, terms, basic requirements, project overview, basic data, physical progress of project construction, flood control standard, flood control scheme, physical progress requirements and measures of flood control, forecast and early warning and emergency response plan, flood control organization and supporting measures, and attached figures and tables.

National Energy Administration of the People's Republic of China is in charge of the administration of this specification. China Renewable Energy Engineering Institute has proposed this specification and is responsible for its routine management. Construction Design Subcommittee of Energy Sector Standardization Technical Committee on Hydropower Investigation and Design is responsible for the explanation of specific technical contents. Comments and suggestions in the implementation of this specification should be addressed to:

China Renewable Energy Engineering Institute
No. 2 Beixiaojie, Liupukang, Xicheng District, Beijing 100120, China

Chief development organizations:

China Renewable Energy Engineering Institute

POWERCHINA Guiyang Engineering Corporation Limited

Participating development organizations:

POWERCHINA Beijing Engineering Corporation Limited

POWERCHINA Northwest Engineering Corporation Limited

POWERCHINA Chengdu Engineering Corporation Limited

POWERCHINA Huadong Engineering Corporation Limited

POWERCHINA Zhongnan Engineering Corporation Limited

POWERCHINA Kunming Engineering Corporation Limited

Chief drafting staff:

| | | | |
|---|---|---|---|
| GUO Yong | HE Wei | YI Chunju | GE Xiaobo |
| ZHAI Zhanghong | LI Xiang | CHEN Nengping | LU Kunhua |
| WANG Xiangyu | ZHENG Qing'an | LU Junmin | WANG Danni |
| WANG Xiaobo | WANG Yongming | WU Wenhong | LI Hongxiang |

Review panel members:

| | | | |
|---|---|---|---|
| WANG Zhongyao | CHANG Zuowei | WAN Wengong | REN Jinming |
| SHI Qingchun | LIAO Jianxin | DAI Zhenfeng | HUANG Tianrun |
| WU Chaoyue | ZHOU Shaohong | ZOU Peng | CUI Jintie |
| ZHOU Wei | YANG Dequan | WANG Wentao | BAI Zilun |
| ZHANG Jinhai | HE Xingyong | LI Shisheng | |

# Contents

| | | |
|---|---|---|
| 1 | General Provisions | 1 |
| 2 | Terms | 2 |
| 3 | Basic Requirements | 3 |
| 4 | Project Overview | 4 |
| 5 | Basic Data | 5 |
| 6 | Physical Progress of Project Construction | 6 |
| 7 | Flood Control Standard | 8 |
| 8 | Flood Control Scheme | 9 |
| 9 | Physical Progress Requirements and Measures of Flood Control | 12 |
| 9.1 | Physical Progress Requirements of Flood Control | 12 |
| 9.2 | Flood Control Measures | 13 |
| 10 | Forecast and Early Warning and Emergency Response Plan | 16 |
| 10.1 | Safety Monitoring | 16 |
| 10.2 | Hydrological Observation and Forecasting | 16 |
| 10.3 | Early Warning | 16 |
| 10.4 | Emergency Response Plan | 16 |
| 11 | Flood Control Organization and Supporting Measures | 17 |
| 11.1 | Organization | 17 |
| 11.2 | Information Management | 17 |
| 11.3 | Supporting Measures | 17 |
| 12 | Attached Figures and Tables | 19 |
| Appendix A | Contents for Preparation of Flood Control Report During Construction Period of Hydropower Projects | 20 |
| Explanation of Wording in This Specification | | 21 |
| List of Quoted Standards | | 22 |

# 1 General Provisions

**1.0.1** This specification is formulated with a view to standardizing the preparation of flood control report during construction period of hydropower projects and specifying the preparation principle, basis, content, and level-of-detail requirements.

**1.0.2** This specification is applicable to the preparation of flood control reports during construction period of hydropower projects.

**1.0.3** The preparation of the flood control scheme during construction of a hydropower project shall consider the actual conditions of the project, meet the safety requirements of flood control during construction, and follow the principles of safety, reliability, technological feasibility, economic rationality, and operability.

**1.0.4** In addition to this specification, the preparation of flood control reports during construction period of hydropower projects shall comply with other current relevant standards of China.

## 2 Terms

**2.0.1** construction period

period from the commencement to the full completion of the construction of a project

**2.0.2** flood control standard

criterion of flood control required for a project during the construction period, generally represented by a flood recurrence interval

**2.0.3** flood control measure

structural or non-structural measure taken for flood control during flood season

# 3 Basic Requirements

**3.0.1** The flood control report during the construction period of a hydropower project shall be prepared each year under the organization of the project owner. The report shall be prepared and submitted to the competent authority before the flood season.

**3.0.2** The flood control report shall include the following:

1 Scope and protected objects of flood control.

2 Physical progress of project construction.

3 Flood control standard for the protected objects.

4 Flood control scheme.

5 Physical progress of the project construction requirements for flood control.

6 Measures for flood control.

7 Flood forecast and early warning, and emergency response plan.

8 Flood control organization and supporting measures.

9 Countermeasures for severe weather, floods exceeding standard, geological hazards, dam/cofferdam break, etc.

**3.0.3** The preparation basis of flood control report during construction period of hydropower projects shall include the following:

1 Current relevant national laws and regulations.

2 Current relevant technical standards of China.

3 Approved feasibility study outcomes and review opinions.

4 Relevant design documents and design changes.

5 Flood control technical requirements for each year.

6 Relevant contract documents of the project.

**3.0.4** The hydrological data and flood discharge capacity of water release structures related to flood control during the construction period shall be reviewed before preparation of the flood control report considering the actual conditions of the project.

**3.0.5** The contents for preparation of flood control report during construction period of hydropower projects should be in accordance with Appendix A of this specification.

# 4 Project Overview

**4.0.1** The project overview in the flood control report during construction period of a hydropower project shall include the general description and construction description of the project.

**4.0.2** The general description of the project shall mainly include the following:

1. Overview of the river basin where the project is located and the status of cascade hydropower developments.
2. Project characteristics, including the location, development purpose, project rank, structure grade, flood criteria, and layout of structures.
3. Construction diversion scheme.
4. Construction duration.
5. Hydropower complex, reservoir area, and downstream protected objects for flood control.

**4.0.3** The construction description of the project shall mainly include the following:

1. Project construction management mode and lotting scheme.
2. Main parties involved in the project construction and their scope of work.
3. Current physical progress of main works, construction milestones, and main construction items and schedule for the flood control year.
4. Implementation of resettlement and special-item works.

# 5  Basic Data

**5.0.1**  Basic data of the flood control report during construction period of a hydropower project shall include hydrological and meteorological conditions, topographical and geological conditions, characteristics of main structures, characteristics of diversion structures, general construction layout, etc.

**5.0.2**  Hydrological and meteorological conditions shall include the following:

1  Characteristic values of main meteorological elements, such as air temperature, precipitation and wind speed.

2  Characteristics of runoff and flood.

3  Design floods, reservoir capacity curve, rating curves at dam site and powerhouse site, rating curves at the inlet and outlet of water release structures, etc.

4  Flood design results of related tributaries and gullies.

**5.0.3**  Topographical and geological conditions shall include the following:

1  Topographical conditions of hydropower complex area.

2  Geological conditions of hydropower complex area and near-dam reservoir banks.

3  Type, distribution, scale, influence range, stability, and assessment of geological hazards of hydropower complex area and near-dam reservoir banks.

**5.0.4**  The characteristics of main structures shall include the layout of water retaining structures and water release structures, and the discharge capacity of water release structures.

**5.0.5**  The characteristics of diversion structures shall include the layout of diversion structures and the discharge capacity of water release structures related to the flood control year.

**5.0.6**  The general construction layout shall include the layout of construction facilities, quarries and borrow areas, temporary stockpile areas, spoil areas, on-site access, camps, gully treatment works, geological hazard control works, etc.

# 6 Physical Progress of Project Construction

**6.0.1** The flood control report during construction period of a hydropower project shall illustrate the current physical progress of the works related to flood control such as hydropower complex, diversion works, construction facilities, quarries and borrow areas, temporary stockpile areas, spoil areas, on-site access, camps, gully treatment works, geological hazards control works, special-item works for reservoir-area resettlement.

**6.0.2** Before the river closure, the physical progress of project construction shall include the following:

1. Physical progress of water retaining structures and water release structures for protecting the construction of diversion structures.
2. Physical progress, completed work quantities, and remaining work quantities of diversion structures.
3. Physical progress of hydropower complex structures under construction.
4. Physical progress of special-item works for reservoir-area resettlement and the implementation of resettlement before river closure.

**6.0.3** At the stage of water retaining by cofferdams, the physical progress of project construction shall include the following:

1. Physical progress, completed work quantities, and remaining work quantities of cofferdams.
2. Physical progress of dam and foundation pit.
3. Physical progress of other main structures.
4. Physical progress of plugging of orifices, adits and access tunnels below the water levels for flood control.

**6.0.4** At the stage of temporary water retaining for dam construction, the physical progress of project construction shall include the following:

1. Physical progress, completed work quantities, and remaining work quantities of dam and foundation pit.
2. Physical progress of dam grouting and the remaining quantities.
3. Physical progress of water release structures for flood control, and the remaining quantities.
4. Physical progress of gate installation for flood control, and the

remaining quantities.

5 Physical progress of other main structures.

6 Physical progress of plugging of orifices, drillholes, adits and access tunnels below the water levels for flood control.

7 Physical progress of geological hazard control works related to flood control safety.

**6.0.5** At the stage from initial impoundment to project construction completion, the physical progress of project construction shall include the following:

1 Physical progress of dam construction and the remaining quantities.

2 Physical progress of dam grouting and the remaining quantities.

3 Physical progress of water release structures for flood control and the remaining quantities.

4 Physical progress of gate installation for flood control and the remaining quantities.

5 Physical progress of plugging of diversion structures and the remaining quantities.

6 Physical progress of other main structures and the remaining quantities.

7 Physical progress of plugging of orifices, drillholes, adits and access tunnels below the water levels for flood control.

8 Physical progress of geological hazard control works related to impoundment and flood control safety.

9 Physical progress of special-item works for reservoir-area resettlement and implementation of resettlement related to each impounding stage.

**6.0.6** The flood control report during construction period of a hydropower project shall illustrate the physical progress and remaining quantities of safety monitoring facilities related to flood control, as well as main monitoring results.

**6.0.7** The flood control report during construction period of a hydropower project shall illustrate the construction and operation of the hydrologic telemetry and forecasting system.

# 7 Flood Control Standard

**7.0.1** The flood control report during the construction period of a hydropower project shall specify the flood control standards, periods and flow rates of the works involved in the flood control year, including the hydropower complex structures, diversion works, construction facilities, quarries and borrow areas, temporary stockpile areas, spoil areas, on-site access, camps, gully treatment works, geological hazard control works, and special-item works of resettlement according to the physical progress and construction schedule of the project.

**7.0.2** Before river closure, the flood control standards, periods and flow rates shall be specified emphatically for diversion works, construction facilities, quarries and borrow areas, temporary stockpile areas, spoil areas, on-site access, camps, gully treatment works, geological hazard control works, and special-item works of reservoir-area resettlement.

**7.0.3** At the stage of water-retaining by cofferdam, the flood control standards, periods and flow rates shall be specified emphatically for the cofferdams, dam and foundation pit, and other main structures.

**7.0.4** At the stage of temporary water-retaining by the uncompleted dam, the flood control standards, periods and flow rates shall be specified emphatically for the dam, foundation pit downstream of the dam, other main structures, and special-item works of reservoir-area resettlement.

**7.0.5** At the stage from initial impoundment to project construction completion, the flood control standards, periods and flow rates shall be specified emphatically for the diversion structures to be plugged, dam, other main structures, and special-item works of reservoir-area resettlement.

**7.0.6** The flood control standards shall be consistent with those specified in the approved feasibility studies.

# 8 Flood Control Scheme

**8.0.1** The flood control report during the construction period of a hydropower project shall specify the flood control schemes of the works involved in the flood control year, including the hydropower complex, diversion works, construction facilities, quarries and borrow areas, temporary stockpile areas, spoil areas, on-site access, camps, gully treatment works, geological hazard control works, special-item works of reservoir-area resettlement, etc.

**8.0.2** The flood control scheme shall mainly describe the layout and characteristics of water-retaining and water release structures, the discharge capacity of water release structures, the flow rates and water levels for flood control, and the safety analysis conclusions of water-retaining and water release structures. The water level of flood control for each protected object shall be determined by flood routing and backwater calculation according to the discharge capacity of the water release structures in the corresponding period.

**8.0.3** Before river closure, the flood control scheme shall mainly include the following:

1 Layout and characteristics of the water-retaining and water release structures to protect the construction of the diversion structures, and the discharge capacity of the water release structures.

2 Flow rates and water levels for flood control during construction of diversion structures.

3 Safety analysis conclusions of the water-retaining and water release structures to protect the construction of the diversion structures.

4 Layout and characteristics of the water-retaining and release structures to protect the camps and the construction facilities, the discharge capacity of the water release structures, and the flow rates and water levels for flood control.

**8.0.4** At the stage of water retaining by cofferdams, the flood control scheme shall mainly include the following:

1 Layout and characteristics of the diversion structures to protect the construction of the hydropower complex structures, and the discharge capacity of water release structures.

2 Flow rates and upstream and downstream water levels for flood control during construction of hydropower complex structures.

3 Safety analysis conclusions and protection scheme of the cofferdams to

protect the construction of the dam and foundation pit.

4　Safety analysis conclusions and protection scheme of the water release structures to protect the construction of the dam and foundation pit.

5　Water-filling and protection scheme for the foundation pit required to pass flood.

6　Safety analysis conclusions of the water-retaining and water release structures to protect the construction of other hydropower complex structures.

7　Requirements for water pumping and dewatering of foundation pits of the hydropower complex structures.

**8.0.5** At the stage of temporary water retaining by uncompleted dam, the flood control scheme shall mainly include the following:

1　Layout and characteristics of the diversion structures involved in the flood control, and the discharge capacity of the water release structures involved in the flood control.

2　Flow rates and upstream and downstream water levels for flood control during construction of hydropower complex structures.

3　Safety analysis conclusions of uncompleted dam for temporary water retaining.

4　Operation mode and safety analysis conclusions of the water release structures to protect the construction of the dam and foundation pit.

5　Safety analysis conclusions of the water-retaining and water release structures to protect the construction of other hydropower complex structures.

6　Requirements for water pumping and dewatering of foundation pits of the hydropower complex structures.

7　Operation mode of the gates involved in the flood control.

**8.0.6** At the stage from initial impoundment to project construction completion, the flood control scheme shall mainly include the following:

1　Layout and characteristics of the diversion structures involved in the flood control, and the discharge capacity of the water release structures involved in the flood control.

2　Flow rates and upstream and downstream water levels for flood control during construction of hydropower complex structures.

3 Initial impoundment scheme.

4 Operation modes of reservoir and gates.

5 Water levels for flood control of the special-item works of reservoir-area resettlement related to impoundment at each stage.

6 Layout and characteristics of the water-retaining and water release structures to protect the plugging of diversion structures, and the discharge capacity of the water release structures.

7 Flow rates and upstream and downstream water levels for flood control during the plugging period of diversion structures.

8 Safety analysis conclusions of water-retaining structures to protect the plugging of diversion structures.

# 9 Physical Progress Requirements and Measures of Flood Control

## 9.1 Physical Progress Requirements of Flood Control

**9.1.1** The flood control report during the construction period of a hydropower project shall specify the physical progress requirements of flood control according to the flood control standards and schemes of the works involved in the flood control year, including the hydropower complex, diversion works, construction facilities, quarries and borrow areas, temporary stockpile areas, spoil areas, on-site access, camps, gully treatment works, geological hazard control works, special-item works of reservoir-area resettlement, etc.

**9.1.2** Before river closure, the physical progress requirements of flood control shall mainly include the following:

1. Physical progress of the water-retaining and water release structures to protect the construction of diversion structures for flood control.
2. Physical progress of diversion structures for flood control.
3. Physical progress of main structures for flood control.
4. Physical progress of special-item works of reservoir-area resettlement for flood control.

**9.1.3** At the stage of water retaining by cofferdams, the physical progress requirements of flood control shall mainly include the following:

1. Physical progress of cofferdams for flood control.
2. Physical progress of hydropower complex structures for flood control.
3. Physical progress of plugging of orifices, drillholes, adits and access tunnels below the water levels for flood control.

**9.1.4** At the stage of temporary water retaining by the uncompleted dam, the physical progress requirements of flood control shall mainly include the following:

1. Elevation of the dam embankment filling or concrete pouring for flood control, the physical progress of the water release structures involved in flood control.
2. Physical progress of dam grouting for flood control.
3. Temperature control criteria for the dam surface to be temporarily overflowed.

4　Physical progress of other hydropower complex structures below the water levels for flood control.

5　Physical progress of installation of the gates involved in flood control.

6　Physical progress of plugging of orifices, drillholes, adits and access tunnels below the water levels for flood control.

7　Physical progress of the geological hazard control works related to the flood control safety.

**9.1.5**　At the stage from the initial impoundment to the project construction completion, the physical progress requirements of flood control shall mainly include the following:

1　Elevation of the dam embankment filling or concrete pouring for flood control, the physical progress of the water release structures involved in flood control.

2　Physical progress of dam grouting for flood control.

3　Physical progress of other hydropower complex structures below the water levels for flood control.

4　Physical progress of plugging structures for flood control.

5　Physical progress of installation of the gates involved in flood control.

6　Physical progress of plugging of orifices, drillholes, adits and access tunnels below the water levels for flood control after the impoundment.

7　Physical progress of the geological hazard control works related to the safety of impoundment and flood control.

8　Physical progress of the relocation and the special-item works of reservoir-area resettlement related to the impoundment at each stage.

**9.1.6**　The flood control report during the construction period of a hydropower project shall specify the physical progress requirements of safety monitoring facilities according to the safety requirements for flood control.

**9.1.7**　The flood control report during the construction period of a hydropower project shall specify the physical progress and operation requirements of the hydrological observation and forecasting system according to the safety requirements for the flood control.

## 9.2　Flood Control Measures

**9.2.1**　The flood control measures shall mainly include the flood control program, flood protection measures, pre-flood inspection and rectifications, inspection and patrol during the flood season.

**9.2.2** The flood control program shall be prepared as follows :

1 Propose the construction schedule related to flood control for the project.

2 Analyze the construction intensity of the works related to flood control for the project, and determine the construction scheme.

3 Develop the construction resource input plan related to flood control for the project.

4 Propose the supporting measures for the construction schedule related to flood control for the project.

**9.2.3** Before river closure, the flood protection measures for the diversion structures during the construction period shall be specified emphatically.

**9.2.4** At the stage of water retaining by cofferdams, the flood control measures shall mainly include the following:

1 Protective measures for cofferdams for safe water retaining.

2 Safety measures for the foundation pit to pass flood.

3 Protective measures for the overflow surfaces of the dam and cofferdams.

4 Protective measures for other hydropower complex structures for flood control.

**9.2.5** At the stage of temporary water retaining by the uncompleted dam, the flood control measures shall mainly include the following:

1 Protective measures for the dam water-retaining surfaces.

2 Protective measures for the embedded parts in the water-passing surfaces of notches and orifices in the dam.

3 Protective measures for other hydropower complex structures for flood control.

**9.2.6** At the stage from initial impoundment to project construction completion, the flood control measures shall mainly include the following:

1 Protective measures for construction safety of plugging structures.

2 Safe operation measures for hydraulic steel structures and equipment of the water release structures involved in the flood control.

3 Protective measures for other hydropower complex structures below

the impoundment level for flood control.

**9.2.7** Pre-flood inspection and rectifications shall mainly include the following:

1 Flood control organization setup, staffing and division of responsibilities.

2 Preparation of flood control equipment and materials.

3 Implementation of project construction schedule.

4 Operation of water-retaining and water release structures.

5 Identification of safety hazards of slopes and geological hazards.

6 Operation of water interception and drainage system of the project.

7 Operation of power, communication, transportation, fire-fighting and other facilities.

8 Evacuation scheme for personnel, equipment and materials to be affected by flood discharging.

**9.2.8** Inspection and patrol during the flood season shall mainly include the following:

1 Inspection and patrol organization, staffing and responsibilities.

2 Working mode of inspection and patrol.

3 Objects and items of inspection and patrol.

4 Frequency of inspection and patrol.

# 10 Forecast and Early Warning and Emergency Response Plan

## 10.1 Safety Monitoring

10.1.1 For the safety monitoring, the monitoring items, content and frequency shall be described.

10.1.2 For the transmission of safety monitoring data, the responsible departments and the monitoring data transmission and acquisition modes shall be defined.

## 10.2 Hydrological Observation and Forecasting

10.2.1 The hydrological observation and forecasting shall mainly include the following:

1. Station network layout of the hydrological observation and forecasting system.
2. Communication mode of the hydrological observation and forecasting system.
3. Forecasting schemes of precipitation and hydrologic regime.
4. Observation and forecasting items and flood reporting frequency.

10.2.2 For the dissemination of the hydrological observation and forecasting information, the responsible departments and the hydrological information dissemination and acquisition modes shall be defined.

## 10.3 Early Warning

10.3.1 Early warnings shall cover precipitation, floods, geological hazards, etc.

10.3.2 The mode, objects and procedure of early warning shall be specified.

10.3.3 Early warnings shall be rated by severity, and the threshold value shall be determined for each level.

## 10.4 Emergency Response Plan

10.4.1 The emergency response plan shall be prepared mainly for severe weather, flood exceeding standard, geological hazard, and dam/cofferdam break.

10.4.2 The preparation of the emergency response plan shall comply with the current national standard GB/T 29639, *Guidelines for Enterprises to Develop Emergency Response Plan for Work Place Accidents*.

# 11 Flood Control Organization and Supporting Measures

## 11.1 Organization

**11.1.1** The structure, participating units and personnel of the flood control organization shall be defined in the flood control report during the construction period of the hydropower project. The participating units shall include all the parties involved in the project construction, and the relevant government departments and other agencies providing flood control coordination may also be included.

**11.1.2** For the flood control organization, the duties and responsibilities of each participating unit shall be defined. If necessary, relevant working groups may be set up, and their respective duties and responsibilities may be defined.

## 11.2 Information Management

**11.2.1** The information management of flood control shall include the collection, transfer, processing and reporting of flood control information.

**11.2.2** The responsible departments of releasing and receiving flood control information and their contact information shall be determined for the transfer of flood control information.

**11.2.3** The workflow, content, time limit, and responsible person for public release of flood control information shall be determined.

## 11.3 Supporting Measures

**11.3.1** The supporting measures for flood control shall cover information communication, power supply, transportation, human resources, materials and equipment, medical care, etc.

**11.3.2** The supporting measures for information communication shall ensure the smooth communication of information and the maintenance of information communication system.

**11.3.3** The supporting measures for power supply shall ensure project power supply and emergency power supply.

**11.3.4** The supporting measures for transportation shall ensure smooth transportation for emergency rescue and evacuation of personnel, equipment, materials, etc.

**11.3.5** The supporting measures for human resources shall include human resources input.

**11.3.6** The supporting measures for materials and equipment shall cover the type, quantities, performance, storage location, transportation, service conditions, etc. of materials and equipment.

**11.3.7** The supporting measures for medical care shall cover the medical institutions and their contact information.

## 12 Attached Figures and Tables

**12.0.1** The attached figures shall include the following:

1. Access roads to the project site.
2. General construction layout of the project.
3. Hydropower complex layout.
4. Construction diversion layout.
5. Distribution of geological hazards in the project area.
6. Emergency escape routes and shelters.

**12.0.2** The attached tables shall include the following:

1. Project characteristics related to flood control.
2. Organizational chart for flood control.

# Appendix A  Contents for Preparation of Flood Control Report During Construction Period of Hydropower Projects

## 1  Project Overview
## 2  Preparation Basis
## 3  Basic Data
### 3.1  Hydrological and Meterological Conditions
### 3.2  Topographical and Geological Conditions
### 3.3  Charactristics of Hydropower Complex Structures
### 3.4  Charactristics of Diversion Structures
### 3.5  General Construction Layout

## 4  Physical Progress of Project Construction
## 5  Flood Control Standard
## 6  Flood Control Scheme
## 7  Physical Progress Requirements and Measures of Flood Control
### 7.1  Physical Progress Requirements of Flood Control
### 7.2  Flood Control Measures

## 8  Forecast and Early Warning and Emergency Response Plan
### 8.1  Safety Monitoring
### 8.2  Hydrological Observation and Forecasting
### 8.3  Early Warning
### 8.4  Emergency Response Plan

## 9  Flood Control Organization and Supporting Measures
### 9.1  Organization
### 9.2  Information Management
### 9.3  Supporting Measures

**Attached Figures**

**Attached Tables**

# Explanation of Wording in This Specification

1. Words used for different degrees of strictness are explained as follows in order to mark the differences in executing the requirements in this specification:

   1) Words denoting a very strict or mandatory requirement:

      "Must" is used for affirmation, "must not" for negation;

   2) Words denoting a strict requirement under normal conditions:

      "Shall" is used for affirmation, "shall not" for negation;

   3) Words denoting a permission of a slight choice or an indication of the most suitable choice when conditions permit

      "Should" is used for affirmation, "should not" for negation;

   4) "May" is used to express the option available, sometimes with the conditional permit.

2. "Shall meet the requirements of…" or "shall comply with…" is used in this specification to indicate that it is necessary to comply with the requirements stipulated in other relative standards and codes.

# List of Quoted Standards

GB/T 29639, *Guidelines for Enterprises to Develop Emergency Response Plan for Work Place Accidents*